Look –

What happens to the scale
When love holds it;

It stops working.

-Kabir

ISBN: 979-8-9935738-4-7 (hardback 8x5), 979-8-9935738-6-1 (paperback), 979-8-9935738-9-2 (ebook), 979-8-9935738-1-6 (audio), 979-8-9935738-0-9 (hardback 8.5x5.5)

Publisher: Mossbridge Institute, 1069 W Broad St #185, Falls Church, VA 22046

Cover Illustration of engaging alien face by Brooks Palmer

Cover design by Cubist, 99designs

First edition 2025

Have a Nice Disclosure!

FOREWORD

Neuroscientist Julia Mossbridge has, on her own admission, written what is a very weird and intriguing book exploring the process of "disclosure." UAP/UFO buffs might anticipate it's a handbook for how to disclose the high strangeness in our skies or the truth that Julia has divined in her groundbreaking work on psychic phenomena and telepathy. But this book is much more than that.

By definition, capital 'D' Disclosure is about finding out the truth about something; But Julia's message is that what is most important is learning how to disclose the truth to ourselves. It's about the journey, not the destination — not the truth we arrive at in the end, if we ever do.

She also makes a disclosure of her own: her growing disquiet about how, when she was a young school pupil in the SOAR gifted child program, she and other children were discreetly tested for psychic abilities; that they were administered unexplained pink drinks and purple so-called 'fluoride' tablets.

What the hell was going on? Julia's personal disclosure moment is her coming to terms with the realisation that she was likely experimented on without her parents' knowledge in a secret CIA/USAF? program investigating ESP in schoolkids.

As Julia laments, we may never know the full truth to disclose but the journey to uncovering it has to be a self-loving odyssey.

Ross Coulthart, 24 Sep 2025

Disclosure 101

WHAT IS DISCLOSURE?

You walk into a room you've walked into at least 300 times before. Maybe it's your bedroom, maybe it's your office space or your kid's room. But this time, an announcement seems to be coming from the walls. Not from any kind of speakers, that would be too normal. It's like the walls are speaking. It's not like that, actually. It *is* that.

They say, "We need to talk. There's something we haven't been telling you. You're not gonna like that we've been keeping it from you, but at the time it felt like the right thing to do. When you're ready to start the disclosure process just say, 'Yes.'"

These are the walls. They are talking to you.

You don't live or work in a smart space with talking walls. You don't own a spaceship, and you're not an extra on Star Trek.

It seems the structure of your reality is shapeshifting into one in which the walls ask you if you want to know information that is probably going to blow your mind. At the same time, the fact the walls are talking is already creeping you out and turning your mind in knots.

You know you're sober. You know you've never had hallucinations of any kind. People actually consider you a bit of a stick-in-the mud with all your pleas to your friends and colleagues — "Don't overreact. Let's see what the science says." Could someone have slipped some kind of psychedelic into your water bottle? Maybe that second-hand pot smoke you keep getting bombarded with now that cannabis is legal is actually making you high?

You're still in this room, so that's good. Not really standing or sitting, you realize you're in a bit of a protective crouch. Like you're stuck in mid-air, halfway to a full squat. You take a moment and notice you need to stretch your legs, your arms. You're not going to say "Yes" until you're ready. Furthermore, you want to know *how* the walls are talking, *whom* you're listening to, and *what it means* to be ready. You have some research to do.

You stretch, you breathe. You go over possible reasons why you might be having a psychotic break. But the walls are silent now.

Maybe you dreamt it? You did sleep a long time last night. And lately you've been having some intense dreams. Maybe this is a sleep disorder, like you're in a waking-dream state. Wait, you can try to work your phone! You can never work your phone well in your dreams. Taking your phone out of your pocket, you read some news. Respond to a few texts from friends. Everything is easy, like normal. You get sucked into doom scrolling and things feel good again, in their doomed way.

"Take your time. We have all the time in the world." The walls are comforting you now.

A shudder goes through you. You feel fear. You also feel hope — that's new. It's small, but it's different. The fear is still there, too. Both are active, swapping places so quickly that they are blending into one another. It's muddy in there, in your feeling flower pot. "Feeling flower pot?"

That's not the kind of language you use. And the walls didn't say it either. Where did that come from? Maybe it's all an auditory hallucination. You're not sleep deprived, but maybe it's stress.

Suddenly, you get an idea — you'll record the walls the next time they speak. That's exactly the kind of thing you shouldn't be able to do if this is a hallucination, if you're in a dream, or if you're involuntarily tripping. You open an app and press record. You wait, triumphantly.

And wait.

And wait.

And wait.

After about 15 minutes you realize the walls are waiting too. You're both waiting for each other. You know, with some kind of clear knowing, that nothing will happen next until you say, "Yes."

You can walk out of the room, of course. It's probably a good idea to do some reality testing. So you walk out of the room and go to the sink in the kitchen. You wash out your water bottle. All the maybe-LSD can go down the drain. You start to sing the Mr. Rogers' song — "You can never go down the drain!" which actually scared you as a kid because you never thought you could fit down the drain in the first place, and then he brought it up as an impossibility — which made you believe it was a possibility.

That's when you notice there is someone else here, washing out their bottle.

Weirdly, instead of wondering who this stranger is, you laugh and say "You can't be too careful, you never know what's in the water!"

The other person looks as pale, sweaty, and nervous as you feel. You are like two little seeds, just beginning to sprout out of your pods.

You recognize this is an uncharacteristic thought, but somehow accurate.

You cringe.

She laughs too hard at your water joke, doesn't look at you, and concentrates too much on her water bottle.

You both have no idea what comes next. And also, eerily, some part of you both does.

The walls say, "Now you two are entangled. Only one of you needs to say yes!"

WHY DISCLOSURE?

You really heard that, just seconds ago. You both heard it. You look straight at each other, in the eyes this time.

You notice the hope and fear wave over her face exactly like it does yours. You are naked with her now, your insides turned outside, her insides are too, as if the walls now speak from inside you both.

Or maybe there are no walls at all. All you know is you are wildly more connected than seconds ago — and scared of what that means.

She speaks first. Quietly. For some reason you're surprised she can speak. You know you can't. You look at her like she's a forest animal who can suddenly talk.

"Are you going to say it? Please don't. Not yet." She pleads and you listen, eyes wide. Of course not. It hadn't occurred to you until now. She could say it too, if she changes her mind. "Yes," she could say, for both of you. Some part of you tries it out. Inside your mind, not under your volition. But you don't say it.

You are about to say something, you can tell, but you don't know what it is. "Why?" That's what you say.

"Because what the hell is this?" She glares at you, very concerned. Then confused about what came out of her mouth. "No. No. I mean why is this happening to us?" That's clearer.

In sync, both of you drink from your water bottles. You don't want to know the answer, and neither does she. You don't even know why you asked the question "Why?" The problem is — to answer the question "Why?" requires a "Yes!" — it means disclosure begins.

You feel stuck in a cage, but somehow aware you've always been stuck. You remind yourself you can leave. You can just walk out. It might be rude to leave, but you don't have to be trapped in this world where you want to know something but also don't want to know anything, and there's someone else who has control over your fate but at the same time you control hers — and none of it makes sense — and your heart talks like the walls.

Suddenly, she puts three fingers into the air and looks intently at you. You meet her gaze, hope and fear in all four of your eyes. There is only one thing to do. She puts two fingers into the air. You nod. Then she holds up one finger.

Your mouth opens. Hers does too. Without trying to, without wanting to, you both hear yourselves say —

"Yes!"

She still has her finger up in the air. Time is frozen. If a stranger walked into the room they would think you were both paralyzed by insight, but at this point strangers aren't even available. There is no other space to not be this space, and no other person to not be you two. There is no sound. There is no time.

Frozen there, you realize your mind still works. Looking into her eyes, you realize her mind is working too. You are sharing insights. No longer needing to be spoken, there are three of them. They don't make any sense, but as you feel the insights, you see them arise in her, and you know she can see them arise in you.

Something catches your attention near the faucet — it's her hand. She's moving her hand. You wonder if you can move yours. You find you can, a little bit. Time has started back up.

You hear sounds on the street outside. She turns off the faucet. You hadn't realized it was even on. It and everything else — except your minds — stopped when spacetime stopped.

So many feelings start to emerge...you're sweaty, your face is wet....

"You're crying," she says.

You are. You are definitely crying. You hear choking, thick sobs. You can't remember why though. It's slipping away. She's fussing with a box of tissues that won't open right, and finally rips the top off. That's funny, you do that too. So frustrating, boxes.

Her kindness touches you. You ask, "Do you remember?"

"I'm trying to. So much detail — it was all real, and clear, and I know it's true. But I feel like it was planted in my mind and then disappeared when time started again. Something about dimensions, beginning and ending, separation and healing, creation, and halves and wholes and bridges...."

She remembers more than you do, but it sounds right. "It was more of a feeling I remember. Like everything exactly true and at home. Like, 'of course.'"

You wipe the counter absent-mindedly with a dish rag. There's nowhere to go now. Nothing will make sense now. It's like your life is over, and a new person has to live the rest of it. You can feel your connection with her blending into your connection with the walls, with the heart-talking part of you. Heart-talking — that's also not a phrase you use.

There is grief for what could have been if you hadn't said Yes. Grief for your old self, who wasn't so naked or open or touched by everything around you. But your new self, this heart-talking self, wants to learn.

She finds a pen and paper in her backpack, and draws three symbols, gives them to you as if she is sending an important message. She copies the same symbols for herself and puts them in her wallet.

"Don't ask me what they mean. I have no idea. But we'll need them."

She walks out of the room, and that is all.

Nope. That is not all. For some reason you run — really run — after her. Hoarsely you shout, "Have a nice disclosure!"

She doesn't turn around. Instead, from down the street, she flips you the bird. You smile.

Nice, HOW?

"Have a nice disclosure?" What's that supposed to mean?

I have no idea why I made you say that. I don't understand what I've already written. I don't know what those symbols mean, but I do know something about their structure. I learned about them when they showed up in my mind when I was in my 7th grade "gifted" class in Libertyville, Illinois in the early 1980s.

I'm sorry to be so cryptic with this story so far, with these symbols, but I feel exactly like you do. I'm not in control here. I know the chapter titles of this book, and I was surprised to start this book writing about you and a woman who were apparently entangled at the sink. I guess I'm on a need-to-know basis with this book. A "just-in-time" methodology is going on, and the time is not just yet.

The symbols came when I was 12, but the title and table of contents of this book came to me 4-5 days ago, and I started writing. Nothing about this is normal for me. When I wrote another very weird book, *The Garden*, I had a plan. I would write a stream of consciousness about whatever showed up in my mind. That was very weird, but it was a plan. This book has even less of a plan.

I also didn't realize I'd be writing in second person ("you") or first person ("I") either. I don't know if I'll ever write in second person again in this book. Or maybe the whole rest of the book will be like that? I also don't have a decent interpretation of how what I wrote in the first two chapters matches the chapter titles. Or how this chapter makes sense.

The feeling I get is that the whole process (this book?) is supposed to give both of us a sense of how disclosure works. It's mechanistically simple — things that were hidden become uncovered. But as it's happening, it's very hard to understand.

That's because the things that were hidden aren't some kinds of magic words that are easily spoken or secret-but-conveyable-in-language insights. They are experiences that just arise in us when we are exposed to this disclosure process.

So I guess that's the thing. Disclosure is a process of uncovering — within ourselves — the truth. The wild thing is, the non-click-bait thing is, that it's not the truths uncovered that are important. Not at all.

What actually is important is learning how to disclose to ourselves the truth. It's the *doing of disclosure* we need to practice. And it doesn't seem to have an end, as far as I can tell.

Okay — that sort of makes sense now. I like that we're learning together, wonderful reader, you and me.

What else do I need to say? Oh — I need to say something about how weird it is to say "Have a nice disclosure!"

How can disclosure be nice? It feels bizarre that 1) a never-ending process that 2) provokes massive change, and 3) is not under anyone's control, and 4) reveals previously hidden information, that 5) makes everything we used to believe uncertain, can 6) feel good in any way.

But it can.

And there you go. Let's have a nice disclosure! Whatever that means!

Love Revolution

No more normal

In middle school I remember learning about the theory of punctuated equilibria. It's the idea that for at least some processes, maybe evolution for one, there are brief periods of rapid change (punctuation) with long periods of stability (equilibria). It seemed like a cool "hurry up and wait" dance move, but I always found this idea suspicious as a theory. It felt like it was a single slice of one perspective on evolution. A very thin, selective slice.

If you look at the shape of a species (morphology), it really does seem to be unchanging for long periods of time. That's how we know that an animal is that animal. For example, something looking like a dolphin split off from something that looked like a pig 11 million years ago.

So, because we identify species across millions of years, we are creating a sense of stability just by looking at the outside of animals. It's sort of a human-perception-created idea — because we think two animals with the same shape are the same type of animal, this creates a feeling of stability in evolution.

The shape is what's available to see on the outside. But what about looking at the inside? And are there also outside changes that are not so obvious?

What could be changing during those times of so-called "equilibrium?" According to the theory of evolution and natural selection, something has to be changing to allow the punctuations of relatively rapid-fire change (for example, a pig-like thing becoming a dolphin-like thing). I keep wondering what kinds of hidden or inside changes need to build up to a create punctuated, rapid change that's obvious from the outside.

As far as I can tell, there are two types of hidden changes — inside and outside — that change all the time during the so-called periods of stasis or equilibrium. These are happening continuously, but at some point they result in what looks like a faster, obvious change on the outside.

During the "normal" equilibrium times, a fish has gills of a certain shape and size, and a pig has hooves with a certain number of "toes." But even during these times, hidden inside-and-outside-changes are interacting, playing, dancing, and experimenting to create a world-wide catalogue of potential creations. And that catalogue of creations is somehow recorded and recalled, just-in-time, on a need-to-know basis, when a punctuated "rapid" change is occurring over a short time scale (a couple of million years or so).

Ack — evolution examples are horrible for this. A couple of million years seems super long.

So what about pencils? They're not naturally-made animals, but maybe human-made tools also go through punctuated equilibria that are actually not so punctuated.

Pencils were pretty much the same for a decent amount of time. (I don't like looking up dates for things because I have an unusual relationship with time and I don't really believe the past or future are well understood.) So let's say that pencils were invented in 11,143 BCE by someone who recognized that burnt wood could leave a mark on a stone. That person quickly learned they could communicate with someone who wasn't there by leaving those marks. Probably within a week that person realized they could keep the charcoal off their hands by wrapping it in a leaf or some moss. So there you have the first pencil.

And the pencil was essentially the same, adding graphite instead of charcoal and putting wood around it at some point, until whenever the printing press was invented (let's say 1443) and the typewriter was invented (let's say 1743).

Those few hundred years between the printing press and the typewriter represent a branch in the evolution of the human-made tool species called the "pencil." One branch became a typewriter — still making marks on things to communicate even today (though almost extinct) — and the other branch remains similar to what we call a pencil (a bit more common than typewriters, but dwindling in numbers). Between about 1943 and 2003, in a relative blink of an eye, the typewriter branch of that tree evolved into a mainframe computer/personal computer/laptop/tablet/mobile phone/smartphone.

What changed during the so-called "equilibrium period" of thousands of years to create such rapid recent shifts? I think it might be the interplay between the inside and the outside of the pencil.

I think around now it's supposed to become clear to both of us what the inside and the outside of anything material actually is, and why from here on out there will be no more "normal." Not just in this book, but in the evolution of tools and humans.

Here is one way to think about it:

But I want to think about what was going on inside the pencil before it broke off to become the smartphone. When I think about "inside the pencil," I'm not thinking about what we call the pencil lead. The inside I'm talking about is not material. It's something about the human need that created the pencil and the human relationship with the resulting tool.

It's like every tool we use is actually a human-tool system. Nothing is just the tool alone. There's a human-pencil system, a human-typewriter system, a human-phone system. And each of those systems is dominated (and sustained maybe) by a human need. The human need is why the system was first invented, and why it changes over time as human needs morph. So "inside" the pencil is actually "inside the human-pencil system."

Maybe what was going on inside the human-pencil system between its invention in 11,143 BCE and the invention of the typewriter in 1743 was that because of the invention of long distance travel and the increase in the literate population, there developed a need for a time-and-space machine like a typewriter (and eventually the smartphone), that would allow people to communicate with more people in very distant places and at very distant times. Pencils could do this okay, but the marks didn't usually last too long and you had to make a lot of them. That was exhausting. You could also just not communicate, but probably there were some things going on outside of the human-pencil system that made not communicating difficult. People needed a time-and-space machine — a way to skip over the boundaries of time and space, to support connection.

Now I'm seeing that the need for a time-and-space machine developed gradually from a series of human-pencil system responses to what was going on outside the human-pencil system. Outside that system, the human-fire, human-wheel, human-weapon, human-farm, human-house, human-clothes, human-boat, and human-rules systems were allowing people to move farther and farther away from each other, kill each other, create money, live longer, and scale confusion in a way that built a need for the time-and-space scaling of communication.

So the human-pencil system, containing the entire history of that system, erupted suddenly after about 8 thousand years to create new but related solutions — the printing press and the typewriter and the smartphone. Those solutions are the punctuations in the evolution of the human-pencil system, all beginning on that fateful day on Oorg 17 of 11,143 BCE (Oorg was maybe like November, let's say).

While there seemed to be a long period of equilibrium or stasis, followed by several punctuated periods of quick change, in fact the inside the human-pencil system was changing the whole time. These changes to the human-pencil system started as soon as the first need arose that was unmet by the pencil itself. This was probably on Oorg 28 of 11,143 BCE, when the inventor of the pencil got her legs chewed off by a roving hungry bear, about which she attempted to warn everyone else by writing the equivalent of "Help! Also, bear!" on a nearby tree.

The moral of the story is that sometimes it takes a long time before the conditions are right to meet even the simplest needs. And once conditions are right, the punctuations multiply and finally the changes going on inside a human-tool system can be seen outside of it. The current blossoming of AI tools is an obvious example of a new punctuation driven by need.

The funny thing is, currently with all human-tool systems we make the same mistake. We think the human-tool systems are more real and essential than the human soul. I know I just sprung the word "soul" on you, so if you want to feel scientific, you can think of the human soul as a unique, naturally made, non-physical, informational system that interacts 1-to-1 with each human body. Without our souls, we can't be fully human, so they are truly essential for us.

That's why "no more normal" is the name of this chapter. No longer can we ignore the human soul.

As we recognize all the systems that are changing on the inside in response to the outside changes, it's easy to see that "equilibrium," "stasis," and "normal" are names for an externally sliced view on human-X systems. Those slices are too small now for a post-disclosure world. The changes in one slice stay static perhaps only from one perspective (the tool part), but not the other (the human part).

It's kind of like peanut shells on the ground. They're evidence that something has changed — the shell of a peanut is evidence of a human-peanut system. Typewriters and phones are like those shells. They are evidence of change, but not the cause of it. They're the shells, not the peanuts. The human-communication system is not the cause either.

The constant unfolding of the human soul is always the cause. The soul also doesn't ever go into stasis. That informational system is always on the move, always changing — always learning goodness. This growing towards goodness may seem horrible in any one slice of time — the learning proceeds in spectacularly nonlinear ways — but it keeps going. And it can no longer be normal to ignore, abuse, neglect, or simply not nurture the growth of the human soul.

There, there's some disclosure.

WHO'S IN CHARGE?

Well, that seems like an important question.

Disclosure begs two questions (at least) — disclosure of what? The other one is — who is disclosing stuff?

The second one is easy to answer, but hard to answer in a way that will be satisfying. It's easy to answer because the answer is simple — you are in charge of disclosure. It's hard to answer in a satisfying way, because you have hidden that fact from yourself. Either you don't think you have any knowledge or wisdom — or you know you have wisdom and you don't understand why you would hide it.

But here you are, reading this book on disclosure that says you are in charge of disclosure. Wouldn't *I* know something about disclosure? Or that stranger you met at your sink? If *you* were in charge of disclosure, why would you be reading this book?

Yet you are *still* reading. So either I am wrong and you are not in charge of disclosure, or I am right that you're in charge, but you have decided to hide the fact that you're in control, for some reason. Or...is there some other option?

Yes, of course there's another option. You think of that one too. Just one more beat.

Not that you know what that means, but what comes to mind when you see these symbols is that maybe your definition of "you" is not quite the same as mine. Or my definition of "you" is not quite the same as yours. Or both. Since you are not me, you are you, it's good to check this out.

Also, by the way, seeing symbols appear before your eyes that magically mean things is probably a sign of being crazy, so there's that, and you're definitely not in charge of that. So get that checked out too, while you're at it.

Assuming you've assessed and hopefully re-assessed your mental health, and found yourself not in any immediate danger, let's work on this "you" thing.

You check with me, the author: "What do YOU mean mean you say YOU?" Well, that's a confusing question. Too many "yous."

Try again. I'll ask you. "What do you mean when you refer to yourself?"

You respond, "My body is one part, but that's not all of it. My memory and life history are other parts, and that's not all of me either. My thoughts, feelings, ideas are part of it — but I am just adding 'my' to things and that's like using the word I am trying to define in the sentence!"

You are right about that.

So you try again.

"Let's assume I am in charge of disclosure. If so, I am disclosing something, which means there must be an active part of me that can hide and disclose things. And it also means there is another wise part of me that knows things that can be hidden or disclosed. This part must be connected to the wider universe of information, or else it wouldn't know things that can be hidden or disclosed.

It also means that before disclosure, that wise part still knows things and that active part is hiding what the wise part knows, maybe for good reasons. It also means that after disclosure, if we ever get around to that, the active part will choose to allow the rest of me (or at least some other parts of me) to learn from the wise part what can be disclosed.

Since the wise part, the active part, and the other parts that can learn from disclosure are all part of me, if I am in charge of disclosure — that means the wise part and the active part are also part of the wider universe of information that can be hidden or disclosed, and that wider universe can choose when to hide or disclose itself. So that means, I am, according to this strange author who asserts I am in charge of disclosure, I am — or I include — all of that."

That was very helpful to both of us. Thank you.

COPING TOOLS

You are not sure why you find this process irritating, but you do. You are also excited. Depressed, strangely joyful, in grief — in shame, angry, delighted, and whimsical. Different days, different ways. You're needing some grounding.

You spend a few days, weeks or months, maybe years just doing normal stuff, not thinking about disclosure, the "I" or what it means to be connected to the universe through the inside of some kind of human-something system.

You walk around. You feed yourself. You let someone you love get you water. You walk around some more. You work.

You lose yourself in the Internet. Or music. Or booze. Or drugs. Or sex. Or food. You make some dumb choices, just to make sure you're still you. And you make some great choices, though you can't help but wonder if that was some kind of cosmic mistake.

These are the disclosure blues. The disclosure blues require a powerful coping tool. That tool is *forgiveness*, which comes with *unconditional love for yourself and others*.

You realize forgiveness is important, it's not hard to conceptualize. You know it means to recognize that we are all struggling at times, and none of us is capable of being perfect, so there must be a way to move on and move forward despite past mistakes. That's forgiveness. You know you need to forgive yourself as much as you need to forgive others. But the hard part is doing it. So you put it on a mental list of things to try, and then you go back to losing yourself. Bad choices, good choices, and all that.

One day you look up and realize — this is the active "hiding" part of yourself. The part that actively hides the information that could come from the wise part. It won't let the wise part disclose. Why can't you just let your active part relax and actively support your wise part in disclosing?

I mean, just do it! Why is it so easy and habitual and maybe even useful to hide stuff from ourselves? What's the magic secret to support continuous disclosure? It sounds ridiculous to even ask. But come on. You are ready.

Okay, the heart-talk is telling us that there actually is a magic secret to continuous disclosure: Forgiveness, which comes from unconditional love. Just the same as the coping tools we talked about seconds ago.

What does that even mean?

Well, unconditional love is like a human response to a natural force we can call Universal Love. One day we will learn the physics of Universal Love. But what it feels like when humans interact with this force is that we are filled with love, and we can love and feel loved without anything needing to change. This is unconditional love. Feeling these feelings leads directly to forgiveness, which comes from recognizing everyone is always doing the best they can, without exception. Even bad people.

Unconditional love doesn't mean you have to hang out, have sex with, or give money to bad people. Actually, the contrary.

If you have the experience of unconditionally loving everyone, including yourself, you will have the natural ability to form boundaries between you and the people you see as bad. They are, after all, doing the best they can. It's just that the best they can isn't good enough to be hanging out with most people, so there has to be some way to protect them and others so they don't do their bad stuff and hurt people. But this does not mean they cannot be loved.

The feelings of unconditional love and forgiveness are not behaviors — there is no consistent outside manifestation of them. They are experiences that are responses to accessing something in the universe that transcends time and space — Universal Love. They are responses to the reality of connection and interdependence. They are the hidden changes on the inside that eventually will erupt to shift the outside realities of an entire community, culture, and world.

Okay, love is important, yes! But let's stay on topic. What does this have to do with continuous disclosure? Or coping tools?

Unconditional love and forgiveness are the tools to help you stop asking this question. That's because loving yourself as you are without needing to change anything (unconditional love for yourself) means realizing that continuous disclosure cannot happen while you are still alive and hanging out in a body.

Because you need to eat, sleep, walk around, poop, talk to people, work, etc., continuous disclosure is impossible. If mystical stuff were always being revealed to your conscious awareness, you would just drool and stare at a wall.

It comes in waves because we can only handle it in waves. Like food, sex, booze, bad choices, even good choices — all of these have to come in waves. These oscillations are signs that there is a natural process going on. Anything that doesn't oscillate is dead.

Accessing unconditional love, experiencing forgiveness for ourselves-as-we-are, forgiveness for not being able to stay alive in a body and constantly receive the hidden wisdom — this is the best coping method for the disclosure blues.

Maybe the reason we have an actively disclosing-and-hiding part is that this kind of pacing is necessary. It's like hunger — the reason you have a fullness-detection system in your brain is that pacing your food intake is necessary for digestion. Food and poop come and go in waves.

If anything that doesn't oscillate is dead — is everything that oscillates alive? That would be interesting. Like the ocean would be alive — or the ocean-land system. Or the moon? Really, the moon-sun-earth system. Is this true?

Now we're thinking back to the target for this chapter — specifically what's the oscillating equivalent of the disclosure process?

In times of disclosure, you enjoy (or not) the emerging wisdom or downloads you are receiving from your wise part that also inhabits the entire universe. But then after that, you integrate that information into your soul, mind, and body — you don't have to think about it all the time, but something has changed. At the same time, you keep wondering what the byproducts are of integrating that information? Are there waste products after integrating wisdom?

There are. They are called dogmas, maybe. Love and forgiveness can ocean-like wash them away, so they can be turned into wisdom.

We are wanting to know how to forgive. How to access unconditional love.

You look at the symbols and don't understand them. And at the same time, you have an idea. What if we asked that actively hiding part to disclose the wisdom of forgiveness? What if we asked each time we needed to cope with the disclosure blues?

"Actively hiding part, please allow me to forgive myself. Please help me access Universal Love. I don't need to understand how, or why it's been hard for me to do this. I just ask that I can do it now. Thanks."

Then you look back at the symbols. You feel more alive. Love is waving around you.

Deep Tech

The rules are wrong, but useful

What if we had to fully describe to another species how a car works, using only colors of light?

Let's say all we have is a light bulb that can be made to shine an infinite number of possible colors, and we also have a room where people will go and experience the light show we create. The light projects all over the room, so there's no directionality to it. There is no change in the light other than the colors and off/on — no dimming and no movement. So we can't show a slideshow or a video of car parts. Also, when people go into the room they don't know they're going to learn anything at all, and certainly they don't know they're going to learn about how a car works. They may not even know why it would be useful to know how a car works, and they may never have seen a car at all.

How would we do this?

My first thought is if we change the color of the bulb to the exact color we want at any time, and the on/off timing is easy to master, we could start out with no light, then choose a white light to alternate with darkness. That might indicate movement. Speed that up in a nonlinear way and it will feel like acceleration. This can indicate the action of the pistons, or it might indicate the view through the windows of a car, depending on how we do it. Shifts in colors and speeds of the on/off signal could differentiate different car components. Maybe. Or we can give up on explaining how a car works and concentrate simply on conveying the usefulness of a car, the environmental effects, or the impacts on human health.

If we did this, what would people experience?

Everyone who comes into that room would experience something different. The success of our plan would depend on how well we anticipated different people's interpretations of the light show. To be really successful, we would need a good sense of how people think and feel. And to do that, we would need to have similar experiences as the people who come in the room.

But we have this!

Just like the human-pencil system, we don't only have the outside, physical stuff at our disposal to do this very difficult task. We don't just have the colors and the on/off switch to play with. We have our own wisdom and creativity, which includes a sense of how people think, perceive, and interpret what they experience. These are powerful tools — our minds!

The human-mind system is just like any other human-tool system — except different. The human mind is connected on the inside to the wisdom of the universe, through our wise part.

Now you think about this.

What you realize is that all the rules about the light switch and the colors are silly, because they admit only to an external connection available for communication. If you can just communicate how a car works by reminding yourself of your connection to the wisdom of the universe, on the inside, you're all set. Your human-mind system sees that if you are connected to the wisdom of the universe, so are other people. And the key to communicating everything (including how a car works, but also actually important things, like how to have peace and live kindly) is to allow the disclosure from the hidden part to continue.

You think about this some more.

You realize your best bet for helping people learn how a car works — and everything else there is to know — is to help them activate their own disclosure engines. As you think about how to do your light show to support that, you reach around in the back of your mind — all the way back to your wise hidden part — and ask this mysterious Universal Love to help everyone know all they can know. And then, that does it.

You recognize now the beauty of the rules of the light-room, the reason they are incomplete, as well as their profound usefulness.

Manifestation?

As your fearless author, I have to say I fear this chapter.

"Matter is mind, so mind matters" is the original title of this chapter. I resisted it so much. I know I have to keep close to the chapter titles that were given to me, so I imagine this one is being said in a snarky, sarcastic tone. Why? Because I was raised in the 1970s in an encounter-group style hippie home where the cult-ish "est" was what you did and *Course in Miracles* and the *Tao of Physics* were what you read.

They were all about how you control your reality and you can get a fancy car or cure cancer by thinking about it. But they also point out that what you really want is enlightenment, which is smashing your ego on a rock. And you can get that too, by thinking about it. As it turned out, none of this worked consistently. So as a result, many people were made miserable when they found out they didn't control their reality. That's why I'm not so fond of the concept of manifestation.

Also, I feel that the idea that you have to watch every thought your mind thinks because it affects the universe is a puritanical hypothesis that eventually tortures everyone because we all have terrifying dreams and horrible thoughts from time to time that we then, through this kind of dogma, decide may actually hurt others, creating an endless loop of self-and-other torture. My idea of Hell is one in which our every conscious thought impacts everything. So when this chapter title showed up, I got pretty pissed off.

But my agreement with my wise hidden part is to address the original chapter titles. Also, there's truth to it. Mind does matter. Where's the truth? And what's the part that makes me cringe?

As I write these questions I realize — this is exactly like the light-room. In the light-room, the lie is that the light show is necessary to communicate. But in fact it's most helpful to spark the realization about the non-necessity and non-reality of the light-room, and the very real direct connection to the universe. Very helpful there.

Going back to the idea of mind mattering, what is the parallel with the light-room?

The mind-mattering idea comes from a truth: Our thoughts and feelings, the contents of our minds, are always connected to the rest of the universe. This is not a one-way street, like we've been talking about so far. It's not just perceiving the wisdom that is revealed during the cycles of disclosure that our actively hiding-and-revealing part offers us. The other way this connection works is as an *active influence* on the rest of the universe.

Here comes more disclosure, for what it's worth.

This active-influence-on-the-universe is not under our ego's conscious control, just like the active-hiding-and-revealing part is not under our ego's conscious control. If either of these were under our ego's conscious control, we'd just have them on all the time and be omniscient and omnipotent, at least for a brief second, before we burned out our bodies and brains.

The active-influence-on-the-universe, like the active-hiding-and-revealing-part, gets a vote about what its influence on the universe is — but not a veto. So even if we could control it with our ego, we would still not be fully in charge of the entire universe. We'd be in charge of our tiny vote.

Both the active-influence-on-the-universe and the active-hiding-and-revealing parts seem to be doing their work from the wise part that contains what some people call our "higher-self" wisdom about what might be best for us. The active-influence-on-the-universe part is outgoing manifestation — it's like the motor system for our whole being. The active-hiding-and-revealing part is incoming intuition — it's like the sensory system for our whole being. Taken together, all of these active parts have been called the *superconscious*.

I think we are more aware of the sensory input than the motor output, which is why it seems so radical to say that our thoughts can affect things. The conscious mind usually has more access to the disclosure that the active-hiding-and-revealing part is providing than it does to the influence that the active-influence-on-the-universe part is sending to the rest of the universe. This seems helpfully designed for our protection, since we rarely know what is actually good for us at any particular time. Like if we didn't have these protections, our conscious anger at someone might just hurt them!

The final thought in this disclosure wave is that the motor and sensory parts of the superconscious need to work together while a person is alive so that each person can be integrated into the universe and also remain an individual body, mind, and soul.

That was a lot. I wonder how much of it is true. The dogma poop thing seems like it's starting up, so I have to remind myself — it's not the so-called "truths" revealed by disclosure. It's the process of disclosure itself.

Back to my concern about this chapter title. A little while ago we talked about what the waste products of integrating wisdom are, I said they are dogmas. They can also be thought of as the terrifying/cruel/evil thoughts and feelings we have at times. In any case, I think these are part of the evidence of disclosure — garbage builds up and has to be taken out.

So in a way the chapter title is inclusive of those thoughts too. Those thoughts "matter" like all trash matters — it has to be taken out, or it will stink up the place.

How do we do that? Taking out the trash is a natural process, and we have freely accessible methods to do it: Dreams, complaining to friends, creative play, being in nature, prayer, meditation, exercise, naps. The late great energy healer and geologist Garrett Walters once sagely told me that you don't have to watch the trash as it's going down the chute. It's less gross that way, and it goes down whether you watch or not.

Having any trash at all means your superconscious is doing its job sensing the wider universe and offering your vote.

I'm starting to think that the deep technology of this chapter is really just about the wisdom to create a habit to take out the garbage. Send it down the chute to compost. In the future, your trash will come up as a flower.

Time Jumping

You wonder what she's doing — is she thinking the same thoughts as you? No, probably not. She's got her own disclosure process and her own trash to take out.

But you find yourself reading about physics and watching videos about space and time, some quantum mechanics stuff. You are amazed and curious and a bit freaked out about how things work in the physical world. It's clear only that it's not understandable, that it's not like you were taught. At first you figured this kind of thing is just as weird as the mind-soul-heart stuff, but more grounded in reality. Now you're not so sure. You notice it keeps coming back to the same place, and that place is impossible to define. It's somewhere between, beyond, inside, and/or below matter, energy, space, time, and mind.

This all means you are in the perfect state to learn about spacetime travel. Not the stuff you already heard about — Einstein's relativity and such. That's more about matter or objects or people traveling through spacetime. This is about *informational* spacetime travel.

You look at that word, informational. Isn't that the same thing? A person traveling backwards in time or to the East in space — how is that not the same as information traveling backwards in time or towards the sunrise? You're right, it is.

But this new technology is about *information-only* spacetime travel. In other words, information traveling through spacetime without any matter or energy carrying it. Information-only spacetime travel is not light zipping around, taking its own time (which of course is really fast). Light is energy that carries information. Photons (light particles/waves), are not information-only, they are information + energy.

Okay, but this disclosure is about information-only *time jumping*. That is the chapter title. This shouldn't be called "spacetime travel" because I am not sure, if there is no matter or energy moving, that information-only stuff is in spacetime or is traveling at all!

From my view of spacetime, information can be anywhere because it really has no position in spacetime without being hooked up to matter or energy. So any movement of information alone really would be less like travel and more like a "jump." Information would jump into and out of spacetime without having to slog through it, like matter and energy must. So that's "time jumping."

What kind of human-tool system does time jumping derive from? It doesn't. Instead, all the human-tool systems probably are derived from time jumping.

Information-only jumping is probably the foundation of every human-tool system that exists. It's not something we make, it's something we use without knowing it. It's what puts ideas in our heads about what we do or what we need to create next.

You wonder if this is veering back toward the mind-stuff. It does seem like that, doesn't it? It seems like that to me too. You ask, how does time jumping work? Information has to be carried by something, yes? If not matter or energy, what?

Anything that must only exist in spacetime, like energy and matter, can be used to carry information *through* spacetime. If information is going to be carried strictly through spacetime, energy or matter or some other spacetime-bound-entity will have to carry it. But that doesn't mean that if information just "shows up" at a particular moment in someone's head that it had to be carried by matter or energy. It might not have moved through spacetime — it might have jumped there from outside of spacetime.

Let's do a thought experiment. You've read *Flatland* by Abbott. You know that living in a particular set of dimensions all your life can make other dimensions seem impossible, and their effects seem arbitrary. You remember the example of a pencil making a dot on the page — for those who live in the two dimensions of the page and can't see or experience the third dimension, what they experience is an edge of darkness appearing out of nowhere.

The thought experiment is this — trace the source of your original ideas or wishes. Notice when something "occurs" to you or when an idea comes "out of the blue." Of course, you won't be able to find the source — it's buried in the non-conscious parts of your mind, which includes your wise hidden part, and may only be revealed by your superconscious (if it wants to reveal it to you).

We talked about how information is revealed by the activity of the superconscious, but how does that information get to your hidden part in the first place? Does it travel through spacetime on a boat, plane, electron, or a photon? Probably not.

If not, according to this very speculative line of thinking, there is probably something about the non-conscious parts of your mind that can access the information-only world. This information is jumping to you — right where and when your mind is — wherever and whenever that is. Where and when is your mind located, anyway? Well, it is certainly not *only* in spacetime, because it can travel way too quickly to the past or future to be bound by the speed of light.

But wait a minute. People can get information about future events that are apparently purely random, it's called *precognition*. It's not always reliable, but it's still statistically and scientifically sound as a real human capacity.

We also know that photons can seem to travel backwards in time — so maybe our rules about spacetime travel are wrong, and we don't have to invoke anything outside of spacetime to explain where ideas come from. Maybe they come from photons slogging backwards through spacetime!

But let's keep thinking about it — if photons are bringing ideas from the future, there are still photons and ideas. These are different quality things. Photons may be able to carry ideas, but they are not equivalent. Ideas "show up" in our minds. Are our minds entirely in spacetime? I think they can't be. Our minds can't be pointed to. Our brains can, but not our minds. They don't have a coordinate in spacetime. You can't point to where your mind is, and "when" your mind "is" is always now, which doesn't make a lot of sense on a spacetime map.

There must be an outside-of-spacetime dimension that is informational-only — your mind, and anything that processes information — has to at least partly live there. That's because information is not equivalent to matter or energy (though matter and energy can carry it), so it doesn't have to play by the spacetime rules. It probably has its own dimension, with its own rules, and these intersect with spacetime rules.

Whatever argument physicists make for why it appears our usual rules for spacetime travel don't fit all of the examples we have — it comes down to that there has to be at least one information-only dimension outside of spacetime that can influence and be influenced by what physically happens in spacetime. It's like an informational substrate of the universe.

Sometimes I call it a "Pervasive Universal Consciousness" — PUC, like the spritely prankster in Midsummer Night's Dream.

You kind of get this already.

But if PUC is "deep technology," you wonder, it must be able to be used for something. It's not just science, right?

You take a moment and think about the Manifestation chapter again. Maybe you re-read it. There are some similarities, aren't there? There must be, for instance, a two-way interaction with the information-only dimension outside of spacetime. It affects us and everything in our world, but we can affect it as well, setting up a beautiful cycle of bi-causation.

But you remember the Flatlanders trapped on a 2-dimensional page, after the pencil has made the dot? Can they influence the pencil?

No and yes. There's no way they can influence the pencil in spacetime. There is nothing they can do in spacetime, because there's no ability to push or change that wall of dark color that just showed up. In order for them to influence the pencil — or more accurately, to influence the being wielding the pencil — they need to use their information-only operative skills. They need to use their intention to ask for the change they hope to see.

If they don't realize another (information-only) dimension exists, they will never be able to affect whomever is wielding the pencil from outside of Flatland. But if they recognize that both they and the being wielding the pencil share *access to and influence on* the information-only dimension, then they might recognize that this is where they can interact.

You wonder, "Is this prayer? Maybe prayer is how we reach out to and learn from every other dimension, because the information-only dimension is interdependent with all dimensions."

You agree that your prayer or your intention, yielded by your superconscious, gives you a vote, not a veto. But if all of the Flatlanders earnestly prayed, then maybe their prayers can make an idea pop into the head of the being wielding the pencil. Maybe that being will draw dotted lines next, to illustrate a Stonehenge effect — essentially communicating with the Flatlanders, with the information-only dimension as a medium.

Look at us! We both realize that this technology is the most powerful technology in the universe.

We're almost right. Except then we have one more realization: it's the only technology in the universe. The entire universe emerges from it.

Have a Nice Disclosure

Love as the Future

You are scared of unconditional love, and that's normal. None of us have been around unconditional love much, even though every major world religion seems to be (at its best) a thought experiment on how to get people to experience more of it.

Here's one of the fears: "If we love ourselves and others unconditionally then everyone will be able to *get away with things* we shouldn't do." The unstated idea is not that we shouldn't ever love, but that *conditional love* and the threat of its withdrawal are what hold society together.

But the thing is, we wildly, over-estimate the power of conditional love. We wrongly believe conditional love is super effective at shaping behavior in the direction we want to see it go, so we fear letting go of that presumed control.

It's like we say, "If I can't wield the threat of withdrawing my love, then I won't be able to have control over myself and others." But when we really look at it, we see that trying to control people by withdrawing love: 1) doesn't work well because it often backfires, and 2) hurts everyone.

Example: Your adolescent kid hasn't cleaned their room for the 10th weekend in a row. They are not getting their allowance, as per your previous agreement with them. But you really want them to clean their room. So you decide not to talk to them or say prayers with them at night until they clean their room. You withdraw your love in hopes it will work. Perhaps you hit them without mercy. Does it work? Maybe. But if it does, it invariably leads to pain and dysfunction.

Your kid realizes that in order to get your love they have to do what you say beyond your agreement about the allowance or other consequences you've given them. They are faced with a choice between bowing to authority versus standing up for what they are trying to do (which is make a choice that is different from yours, pay the agreed-upon consequences, and differentiate themselves from you).

Without unconditional love emerging at some point to heal the situation, there will be insecurity in relationships and bad behavior the rest of our lives. It's like we have to turn around to face the past, and apply some unconditional love to those times when we weren't able to experience it, through no fault of our own.

Example: You overate again. You're stressed out, and you are using eating as a coping tool. So you decide to withdraw love. You berate yourself. Does this make you stop using eating as a coping tool? No, you value yourself less and eat even more poorly. You try to shape your behavior the way we were all taught, with so-called "love" conditioned on perfection, and you hurt yourself. You also hurt the people who are trying to love you, who don't understand why you are feeling so down and treating yourself so poorly.

Even reading about these examples makes me want to apply Time Travel Therapy to weave love back in time to when each of us experienced these kinds of things. Maybe it was just yesterday, or today, or it's right now.

Some psychologists in the Israeli army spearheaded the use of Time Travel Therapy as a healing tool for soldiers with PTSD. The idea is that if you have lived through something very difficult (which is more common than otherwise, the older you get), you have wisdom that you didn't have back then. Those psychologists helped soldiers imagine going back in time to their earlier selves who were going through the hard thing. They offered love and wisdom to their past selves.

You try this.

When you try this, you tell your past self that you will thrive — that you are loved through spacetime and beyond.

You even give yourself tips about what to do next. This kind of treatment worked for the soldiers, and many other people. Now it is working for you.

Why does it work?

The crucial healing piece of Time Travel Therapy is Universal Love, woven through time like a stitch that heals a wound. If you went back in time to berate yourself further or tell yourself you should have done better, that's not helpful. Our human experience of Universal Love — unconditional love — connects us through time. We become our own space-and-time machines. We carry unconditional love through the informational dimension, time jumping to heal ourselves then, now, and in the future.

Once you recognize the power of healing through time, it spreads out over space to bring love to others, like a love bubble. There is no othering — everyone is contained in this love bubble. Once you see that you were always doing the best you could at every moment given the resources you had, this love spreads out through time to your future. There is no othering — every time is contained in this love bubble. You are whole in space and time.

But the fear comes up again. If there is no othering, if everyone is loved unconditionally, does that include pedophiles? Genociders? Murderers?

Yes.

Unconditional love supports boundaries and consequences, because boundaries and consequences don't mean the loss of love.

Someone who hurts people will have consequences *even while they are loved unconditionally*. Those consequences can potentially be rehabilitative in this new framework, because they no longer live in a framework that requires anything to change before love can be doled out. Instead, the feeling of unconditional love is always available — and can be accessed by anyone via Universal Love.

Instead of making people feel like they can get away with bad things, unconditional love ends up supporting change.

You're right to notice this seems paradoxical. Unconditional love means loving without anything needing to change — but it magically motivates change toward the good, without requiring it. In the meantime, as the person is changing (or not), they can be removed from the rest of society if that's what's important to do to unconditionally love them, ourselves, and others.

Why is this chapter called "Love as the future?" I just now realized it's a double entendre. It means: 1) think of the future as love, and 2) we can try to love like the future loves us. Both are unconditional.

Now you realize clearly, like I just did. The future *is* us, with full knowledge of our wrongdoings and misperceptions and weaknesses, with the wisdom to love them all.

A PHYSICS OF LOVE

I don't understand the physics of love, but I think there should be one. I suspect I was made to write this chapter to push myself to say something about what I think a physics of love would be like.

The most fundamental aspect of a physics of love would have to be that love is something that physicists should care about even more than psychologists. Universal Love is more like a natural force — like electromagnetism or gravity — than like a feeling.

The feeling of unconditional love is real, in the way all emotions are real, but it's not physical. If unconditional love is the feeling we get in response to Universal Love, then the brain states related to the feeling of unconditional love are physical. But all feelings themselves (aside from the brain states related to them) are not physical. They exist in information-space. They are in the mind, not the brain.

I feel suddenly apologetic tossing these ideas around without testing every one of them, but it's a bit late for that. If I'm going to apologize, page one should have been an apology for wild speculation as well as for drawing on theories from around the world across spacetime that I don't even know I'm drawing from.

That apology is not forthcoming.

Moving on then. It definitely seems like Universal Love is physical. But I don't like what I said before, about it being like electromagnetism or gravity. It's not really like them — they are more like it.

Maybe Universal Love has many expressions, and everything physical is an expression of it.

For this to make sense, we need a physical definition of Universal Love. I think the physical definition is — "that which connects." If that's our definition, then the four fundamental forces — gravity, electromagnetism, and the strong and weak nuclear forces — all are physical expressions of Universal Love.

Gravity connects masses to one another, electromagnetism connects masses/energies and fields, and the strong and weak nuclear forces connect atomic possibilities. But there are more expressions of that which connects — spacetime connects information about everything that's physical/energetic, so spacetime is an expression of Universal Love.

What about nonphysical things? Is there something that connects pieces of information to each other in information-space? It seems like it. When we get an idea or think of a person, we also get related ideas and people. They are seemingly already connected in information-space. By what? What has the ability to connect informational elements in information-space? I call the thing that has this information-connection ability a "mind." So as information-connectors, minds are another expression of Universal Love.

Yes, I hear you asking and answering this with me — is there something that connects information to minds so minds can connect different informational elements? There has to be, or time jumping — which we talked about as the source of all ideas that come "out of the blue" — can't happen.

As I wonder about this, I keep thinking about bosonic particles — these include photons — which are so different from fermionic particles. Fermionic particles, like electrons, protons and neutrons, behave so much more like what we are used to seeing in the macro-level physical world. For instance, two identical fermionic particles cannot be in the same place at the same time. They don't overlap in spacetime without having different spins. But the standard model of physics says that two identical bosonic particles (like two photons, two gluons, or two Higgs bosons, for instance) can be in the same place at the same time without having different spins. In other words, they can have the same quantum states in the same spacetime coordinates.

It sure seems like this means they have some informational components (their quantum state configurations) that are not like how we think of physical things. So I guess bosons are my bet for expressions of Universal Love that connect information-space with physical/energetic components of minds, like brains or computers.

You are asking if there is any physical force, field, substance, or process that doesn't relate to what we've already covered? Probably we'll know in the future. But everything in physics that has ever been examined is really all about relationships. All of the forces and capacities of particles are described in relationship to one another. It seems to me right now that all of physics is actually about "that which connects."

So I guess a physics of love might be just that — physics, with the acknowledgement that everything physicists study is an expression of Universal Love.

Welcome to the Inspiracy

You're stretching your back. It's been a very weird day. Week. Month. Year.

You thought there would be some more information about UFOs, aliens, government plots, and psychic stuff in this book. But you also kind of like that no conspiracies are being foisted upon you. There are enough on YouTube already. Some seem possible, others less so. What seems impossible is distinguishing truth from fiction.

Sick of conspiracy, you are beginning to recognize that there's something new growing inside of you. Like an internal conspiracy arcing towards truth, goodness, grace, love. It feels like technological advancement at the highest level — at the level of your soul. The opposite of conspiracy, it's an "Inspiracy." You're tired of waiting — you've read this book and done a million other things to try to understand. It feels like you've waited your whole life for someone to just say the truth. And you feel you'll know it when you hear it.

"Will you know it when you say it?," she asks as she walks in.

(Where did she come from?)

You don't talk for a minute. It's hard to believe she's here again...with the weirdness you shared back at the beginning, with her flipping you the bird...you figured she might have been a hallucination anyway. Or not. But you just thought she wouldn't be back. Not in this lifetime.

You decide to speak first.

"Did you learn anything about — you know — aliens?"

She smiles.

"A lot, actually."

You wait as she sits down on the couch next to you.

She begins to speak, but not with her mouth. It's like the walls are talking all over again, but without sound.

Your heart is receiving her words — but they are actual, technically interesting sentences — not just the vague but pungent heart-talk you have come to expect. Actually, you realize, you don't know what to expect from your heart anymore. Your heart has shown itself to be capable of accessing Universal Love and time traveling, so why not some telepathy with your favorite mysterious stranger?

You feel a sense of inner knowing that predates all external knowledge, like you're both babies basking in the universe's wisdom and soaking it up like the sun. You feel if everyone had this experience, lots of problems could be solved. No more need to try to wrest control from others who also don't have it, for instance.

"What about you?" she asks.

"I think I don't know anything anymore," you say, smiling.

You reach for her hand exactly at the same moment she reaches for yours.

Not The End.

From Me to You

SOAR GATE

One of my favorite expressions is "Sunlight is the best disinfectant." I like to think about how, when you're walking late at night and you might be afraid, the best place to walk is in a well-lit area. In other words, *being seen* is the best protection. And yet — it can take decades to feel safe enough to walk under the streetlights. The risks of shame, blame, stigma when becoming non-anonymous are real. So if you notice the tone change between this section and the rest of the book, that's where that comes from. I'm about to be as careful as I can about some personal disclosures, and I am hoping I do not hurt anyone in the process, including myself.

Writing and editing this book has had an effect on me that I didn't anticipate. I wasn't originally going to include any personal disclosure, only science/tech stuff and experiential spiritual material. But as I write this, I'm listening to the Sept 2025 press conference for the survivors of Jeffrey Epstein's predation and abuse, and I'm reading Pulitzer prize-winning reporter Kate Woodsome's powerful words about trauma, shame, and healing. The bravery of the survivors brings me to tears. Even though nothing that I will disclose here compares to their experience, I am still buoyed by their strength. The truth has a way of making its own wisdom, and here I'll offer some that it has made in me.

I deliberated and asked several friends about this section of the book. That's because I have a security clearance, and there is of course a fear of losing that. But I haven't been "read into" anything — I have not been told any classified secrets. In the strange world of clearances, that means there may be things I write in this book that happen to be classified, but I don't currently know that they are classified, so I can talk about them with impunity. They are all based on my own opinions, experiences, and/or non-classified public records.

Interestingly, if my clearance ever becomes active, I understand I will then *not* be able to say these same things if I learn they are part of a classified program. The rules of secrets are bizarre, arcane, and fascinating. They reveal different approaches to the universal struggle to cope with complexity, lack of control, and fear — more on that in the next chapter.

Ultimately I decided that I want to do my part to support the truth-telling and forgiveness process that I believe is emerging worldwide, and demonstrate the effect that writing about disclosure has had on me personally. When we combine our truths with awareness that each of our lives is meaningful, I suspect we will begin to end structures and processes that support unnecessary and ultimately harmful secrecy.

Disclosing my truths starts with a Reddit post response I wrote in Summer 2025, while I was editing this book. The original post was about a gifted-students program called "SOAR." The original poster spoke about being part of this program in the early 90s, stating it sounded similar to the controversial GATE (Gifted and Talented Education) program. It's probably a coincidence that SOAR GATE is one letter off from STAR GATE (the CIA-launched program to create psychic spies), but the echo definitely rings true.

As most memorably detailed by "Merciful Martin" (recently revealed to be Jordan Jozak), some students in the GATE program in the 90s and 00s remember being tested for psychic abilities, playing with Zener cards, taken through out-of-body experience training, performing hearing and cognition tests, and drinking a thick pink or green liquid that they were told was fluoride or an antibiotic.

I was in the SOAR program in Illinois in 7th grade and prior to that I was in special "set-aside" reading and math programs for the "smart" kids. When I read the original post on Reddit, I was surprised to hear SOAR existed beyond my school, and also that it continued into the 90s, apparently as GATE. SOAR hasn't gotten as much exposure as GATE, so I am including my Reddit reply here as a starting point for my own personal disclosure.

⊗ ⊙ ⊡

"I had the same experience but in the early 80s in 7th grade at Butterfield School in Libertyville, Illinois. We had pull-outs for 1:1 sessions that I don't remember (but dreaded), Zener card playtime, weird hearing tests (like way too many tests for students who were not at risk for hearing loss, using super high-pitched tones), field trips, constant cognition testing, oh yeah — the weird obsession with fluoride drinks and tablets. My dad told me fluoride was for teeth, not stomachs, so I don't remember drinking the pink drink — maybe I just spit it out?

Most of our work in class I don't remember, but I think we kind of did our own thing. Our teacher was Mrs. Thomas — I initially misremembered her name as Mrs. Larson, but I realize I was confusing my teacher with my guidance counselor. Pat Larson is mentioned in a *Chicago Tribune* article about SOAR from the late 1980s (*To be or not to be? Pupils in gifted classes know.* July 28, 1988, updated August 8, 2021. No author listed). That article says the program came to the schools in 1983. I was in it from 1981-1982 (the first year of the program in our schools), so this was either a newspaper mistake or we were in a pilot year.

There were about 10 kids in our class. When I asked my mom about it in early 2024 (specifically whether they had been asked for parental consent for SOAR) she said, "Nah. They were studying you. They did what they wanted." I said, "Who were *they*?" She said, "I don't know. CIA? I have no idea. They didn't even tell us you were in the program. You did." She's 86 and her memory of events in her life is still pretty good — but during that time she was going through a rough patch in her life, so it's possible she forgot about being asked permission? I suspect not. At that time, research studies on "gifted" children were considered an honor for parents.

I have been doing some sleuthing about the program and found a few sources for the weirdly creepy name "Students On Active Research." Here they are:

1. Obituary of Joyce P. Ross (b. April 17, 1930), who apparently started the program in Aiken, SC. Her husband was Chuck Ross, a chemical engineer at DuPont, which became Savannah River National Lab.

2. One of the earliest descriptions I could find about SOAR and its creators (*Intellectual Flight Offered SOAR Program Students.* Lynne Katonak, *Aiken Standard* Nov 21, 1978, p. 5.).

3. About 100 other Aiken-area results when searching for the exact phrase "Students On Active Research" on newspaperarchive.co m.

4. A 1992 document stamped by the Department of Education that looks like a practicum report (*Improving the Writing Skills for Third, Fourth, and Fifth Grade Gifted Students through the Use of Developmental Processes.* Mary Helen Freeman. 1992 dissertation).

There's a notable lack of search results in any other county or town in the country — but you can find those in your area by searching for "SOAR" with a different version of the acronym — in my case, "Selected Opportunities in the Academic Realm." Here's one from the area I grew up in (*Program Helping Student Needs.* Susan Ewait, unnamed Chicago-area newspaper, Sept. 19, 1987, p.75).

It seems to me likely that SOAR was a predecessor to GATE for kids in the 80s/early 90s. I don't know if government was involved at all, but if the federal government beyond Department of Education was involved, the name strikes me as likely as an Air Force program rather than CIA (SOARing flight). In any case, Aiken, SC is near Fort Jackson, SC but also Augusta, GA — which is where Fort Eisenhower is (now home to the US Army Cyber Command and the Cyber Center of Excellence). And it's also near Savannah River National Lab. So it's not like there is no military or intelligence influence in that area.

Where I grew up in Libertyville (a suburb of Chicago), the closest military installation was Great Lakes Naval Training Base — but my father worked for the Dept. of Energy as a contractor at Argonne National Labs, so that was nearby as well. After investigating a bit more, it appears our teacher Mrs. Thomas's husband was a drill sergeant at Fort Jackson and moved to Libertyville in 1969. Could easily be coincidence, but that was the timeframe (early 70s) that the program began in Aiken.

My overall impression is my mom was right. The SOAR students were being studied — whether it was because many of our parents were federal workers, because some of us had seen UFOs, had weird psychic experiences, or had super-great test scores — I don't know. I do know that when I requested my academic record from Butterfield, all sections were redacted that contained any teacher comments (which is where SOAR notes would be) through all 8 years. I know this because I have a hard copy of my first-grade report card, and can see the comments there that were redacted later.

To make myself less paranoid, I would love to ask others in (and out) of the SOAR program to request their transcripts and see if they are redacted. A friend mentioned it's possible everyone's might be redacted in Illinois because of a law about redacting comments on report cards from the 70s and 80s — but I can't find mention of that law. Also, redaction takes a lot of work, so it seems strange they would do it for everyone who went through the school system.

Anyway, if you were in one of these programs, consider paying the 15 bucks to get your transcript off microfilm. See if there is any mention of the program. Or if you're lucky and have a paper copy of your report card — go take a look. In general, let's assume positive intent and try to understand more. My experience wasn't all bad, but it was very weird."

To reassure myself and you, I'll say that every person named in that Reddit post is already dead, and the ones I knew (Mrs. Thomas and Mrs. Larson) seemed to have the best intentions at heart. They were kind, good people. They let us write books in class! I wrote one about a girl who had one personality while awake and another while asleep, as well as a weird surreal novel about characters who all came together at the end. All of this in the context of the incessant and bizarre testing. If anything nefarious was going on, I believe this means these teachers were caught up in a secretive program with multiple goals that were not shared with them.

Here are two versions of my transcript from 1st grade — the actual report card kept by my mother (on the left), and the electronic copy of the redacted version sent to me in 2021 (on the right). The electronic versions of my 1st-8th grade transcripts sent to me in 2021 look like the one on the right — that is to say, all comments are redacted, including the comments about the SOAR program from 7th grade. I only have a hard copy of the 1st grade report card.

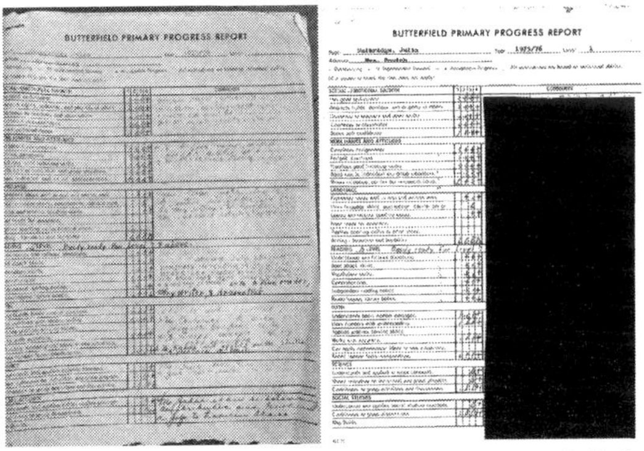
BUTTERFIELD PRIMARY PROGRESS REPORT

BUTTERFIELD PRIMARY PROGRESS REPORT

[redacted]

Two versions of my first grade report card (left, hard copy from 1976; right, redacted electronic copy from 2021).

Selective memory loss.

Beyond not being able to remember whatever I dreaded occurring in that small room on a weekly basis, I have a decent and sometimes photographic memory for many other life events — especially those that were emotionally charged. But I experienced one other time of unexplained profound memory loss for the events of an apparently emotional day. It somehow seems to me to be related to the SOAR surveillance/testing program, yet as always, I'm not sure about that.

During college, I took 6 months off from school to try to understand the direction of my life. I had enough AP credits and could still graduate with my classmates, so I moved from Oberlin, Ohio to Palo Alto, California to live with a friend who had already graduated. I needed a job to pay the rent, so I believe I looked in the classifieds and discovered that a government contractor in Palo Alto was looking for an administrative assistant who had some computer/word processing experience. Another memory is that my dad gave me a contact at Lockheed Martin. But when I asked him about this recently he said he may have had a contact there, but he doesn't even remember me taking time off of school — so I am not sure about that.

To my surprise, the company invited me to interview and I started the job immediately, that day.

The first thing I remember is talking with a man behind a desk for the interview that morning. I sat across from him. I still have a little glimmer of an equipment area or lab behind me, but my focus was clearly forward, toward my new boss. I remember his desk was in front of a window, and it must have been on the first floor, because I could see the parking lot through the window behind him. I felt something was off or weird about the equipment behind me, but that's all.

The next thing I remembered was the end of the day. I was typing a letter on a word processor. I realized I was crying as I typed the letter, and I saw that my hands were shaking. Neither crying in public nor shaking hands were things I had experienced before in my 20 years. I still don't know why I was so upset. At the time, I told myself it was because this was really a missile company, and I was a peacenik.

I remember reading the letter and having the odd experience of realizing I was typing a resignation letter, but I am not sure if this was a metaphor. In other words, it could have been a business letter that I was typing for my boss, but as I was typing it, I felt I needed to quit the job. It is unclear. Regardless, this letter experience was bizarre partly because it was a surprise to me what I had been typing — as if I had woken up in mid-activity. Also, it was strange because I could only remember the first 30 minutes of that day — my interview. So the day I remembered hardly required a formal resignation.

I remember telling my boss I was quitting. He expressed that he thought I could be a great asset, but I told him I couldn't deal with the missile-making aspect of the company's work.

I still don't know how to explain the whole experience. But after talking with several people who were in the GATE program, it appears this kind of selective memory loss for emotionally charged (but not recalled) experiences is a common thread among GATE-program folks with otherwise above-average memory capacity.

Genetics and cognition.

Around 2018 or so (in my late 40s) I enrolled in a cognition study on 23andMe at the website's suggestion. I bring this up because my experience in that online study suggests that I was being studied beyond my time at SOAR. The nature of the study was oddly familiar: a longitudinal, 18-month study in which they'd present me with cognitive tests, ask me what I'd been eating, and inquire about my medical history. Now as a mid-career adult cognitive neuroscientist, I was keenly interested to see what tasks they'd have me do.

When I performed the cognitive tasks, I realized all of them could be "dual-use" tasks — in other words, they could easily test for both cognitive and psychic capacities For instance, there was a balloon-popping task. You were supposed to click a button that, with each click, would fill up a cartoon hot-air balloon with gas. At some seemingly random point the balloon would burst — you got more points the bigger it got without bursting, but if you kept filling it until it burst you lost a lot of points.

This kind of task can be used to look at risk aversion as a personality trait and gambler's fallacy behavior (the belief that if you lose a few times in a row, next time you'll win). But also, if someone is really good at stopping the button-pressing before the balloon bursts (but always pressing it the most they can before it bursts), they could be precognitively predicting the random burst times. So this was a great precognition-screening task as well. I think there were three of these well-done, potentially dual-use tasks. Psychic testing was not disclosed in the consent form, as far as I remember, so if these data were being gathered without consent, this would be unethical human testing. Again.

Clearly the researchers were trying to link genetics, medical history, and cognition — a useful thing to do in a world moving toward individualized medicine and CRISPR technology. But there were two very odd medical questions that could not be derived from my genetics, eating patterns, or cognitive data. One question was about a ganglion cyst on my ankle. I was diagnosed with it in college. The question was very personalized — they literally asked, "Were you diagnosed with a ganglion cyst on your ankle in your early 20s?" The answer was yes, but it really shocked me, because I knew that this level of medical information was not extractable from any other information I told them, nor from my DNA. I suspected they must have had my college medical records, because I know there is no working hypothesis that people with such-and-such a medical profile often have ganglion cysts on their ankles. So this was 95% likely to be a question they wrote just for me.

The other question that felt very strange was, "In the past week, were you diagnosed with a torn rotator cuff?" I hadn't been diagnosed, but that's just because I didn't go to the doctor with my shoulder pain. I knew that I had a torn rotator cuff, I didn't know how it happened, and I intended to just let it heal because I knew the doctor couldn't do anything about it. The researchers could not have known about it even if they had my medical records, my DNA and my cognitive test results.

I know this is very unusual, and it's the kind of thing that can make you a little paranoid about surveillance. Or make you start to believe in telepathy.

During this period of time I had a dental appointment, and the dentist gave me some kind of pink fluoride rinse, similar to the kind I had as a kid (but not viscous). I washed my mouth with it, but afterwards my skin turned bright red like a radiation burn. I had never had this reaction to a fluoride rinse before. They thought it was an allergic reaction, and it went away 20 minutes later. The same thing had happened to me a week before, when I was just sitting at home, writing a scientific article. My skin turned bright red on my face, neck and arms — not as a result of a pink drink or anything that I was aware of doing, drinking, or eating. Twenty minutes later it disappeared entirely. I checked in with a friend who studies UAP exposure effects on humans, and he said it sounded like I was exposed to ionizing radiation somehow.

Before 23andMe went bankrupt, I deleted my account but I was told that my genetic data would still be available to any researchers who had requested it for their study. I specifically asked about bioterrorism concerns around any foreign group having access to DNA linked to cognitive results from US citizens, and I suggested that cognitive testing data and genetic data should be uncoupled, but they did not agree. The whole episode makes me think that someone with more access should look carefully into how these data are being handled today.

Psychic research and secrets.

It is publicly known that psychic research was performed on children across the US in the 50s-80s (CIA Reading Room C02944291, LN734-35, p.75 onward). Much of this research appears to have been done without consent. My mother remembers no consent being asked of her or my father for me to do any of the SOAR activities. This included in-class meditations (which I thought were calming), 1:1 meetings with different counselors (which I still can't remember, but do remember every part of the path to and from a claustrophobic meeting room and dreading whatever went on in there), psychic testing, weird hearing tests with strange tones unlike normal hearing tests, field trips to enrichment events, IQ/cognition testing at the school and elsewhere, and pink drinks/purple fluoride tablets. Some people involved in the more SOAR-like aspects of GATE also report no parental consent being asked, and describe their own dread about 1:1 sessions they don't remember. As a result of all this, on 16 November 2024 I filed a FOIA request with the Air Force to ask about SOAR. It began, "Any information you have about me related to my screening/testing/training/development in any Air Force-related programs from 1973 to present day (2024) be released to me immediately," and it went on to explain some of what I've already described about the SOAR program and follow-up testing.

A wild and rapidly unfolding series of events emerged after I made the request. I sent the FOIA request on Nov 16, 2024. On November 20th, I was contacted by a federal recruiter for a job I'd applied for many months prior (in February 2024). On November 21, I completed the first interview and began the second-level application process for that job. On November 22, I received an email from the FOIA office asking me if I wanted to be more specific about my request.

⊗ ⊙ ⊡

"In order for our office to continue to process your request, we need clarification into your affiliation with the U.S. Air Force. Were you Active Duty, Reservist, Air National Guard (ANG) or affiliated in some other capacity? Additionally, what were the dates of your affiliation or your discharge date? Additionally, what offices or programs within the Air Force do you believe have records on you?

For the records you are seeking, requests should provide a description that define the specific records you are requesting in enough detail so that they can be located with a reasonable amount of effort. Generally, a record is reasonably described when the description contains sufficient file-related information (type of document, title, subject area, date of creation, originator, etc.); or the request contains enough event-related information (date and circumstances surrounding the event the record covers) to permit the conduct of an organized, non-random search. Please provide the names of the records or specific descriptions of the records you are seeking information on.

Our office will place your request on hold, pending your response, until 5 December 2024. If we do not hear back from you by that time, we will administratively close your request. Please note that without this necessary clarification, our office will be unable to process your request.

We have assigned number XXXX-XXXXX-X to identify your request; should you need to contact us about your request. Please direct your questions to XXXX@us.af.mil and use this assigned number to assist us in responding more promptly.

Very Respectfully,

HAF FOIA TEAM
SAF/AAII (FOIA)"

⊗ ⊙ ⊡

This response was unusual, because very little time had passed since I'd sent the request (four days!), and FOIA is not known for its responsiveness.

I said yes to dropping the FOIA request, since I didn't know exactly who administered the SOAR program, or even if any governmental department or agency actually did. Also, in the back of my mind it occurred to me that my new potential employer might not want an outstanding FOIA request in my name.

Not more than three minutes after I sent my "please drop the request" response, I got the call saying I'd passed the first interview and now would receive access to the cognitive assessment (the next hoop to jump through for the government job). I know I received this call within three minutes, because I wrote in my journal that I'd decided to dismiss the request, and then on the next line, I wrote "Got assessment within 3 minutes!" Whether this remarkable string of events was a coincidence or not, I did end up passing all assessments and jumped through all the other hoops they had, receiving the job offer in early 2025.

This offer was rescinded in the DOGE flurry of March 2025, so I can't look into any more details about the connection between FOIA and this particular job application process. I was left wondering whether the job recruitment effort was just another attempt to follow up on my cognitive status, or whether I am being paranoid and/or grandiose to even consider this was another surveillance attempt.

It turns out the answer doesn't matter. The key thing is to assume that my well-documented experiences are real, and to move forward with the mission of trying to do good things in the world.

Studying the effects of radiation?

In the past year, I did some genealogy research and found my mother's father's draft card. I discovered my grandfather worked at the Shattuck Chemical company, a company that mined and processed uranium in a plant near Denver, where my mom's family lived in the 1940s. It became a Superfund cleanup site, but of course not during the time my mother would have been exposed to uranium dust. When I told her about this, my mother revealed that both her parents had worked there, but her mother was an office worker so she had less exposure.

Some background on my parents might be important here. My mother was given a full-ride scholarship to University of Chicago, and she told me one of the psychology faculty (a Dr. Fraser or Dr. Frazier) was really into talking about putting "medicine" in the water to cure mental illness. She didn't know what medicine he was talking about, but this was in the mid-50s so it was during the time of MK-ULTRA, the illegal and unethical CIA program that, among other things, dosed unsuspecting people with LSD.

At one point, Dr. Frazier/Fraser took my mother aside and told her the faculty had taken a chance on her, given that she was from such a poor, uneducated family and was the first of her family to go to college.

I try not to be paranoid, and I like to assume that this opportunity was really about her brilliance, which is irrefutably real. But of course University of Chicago was deeply involved in the Manhattan Project, which had been successful — so it makes sense that the federal government would continue working with U of Chicago on any related projects, including studying the capacities of individuals who had been exposed to radiation as children.

Meanwhile, after getting his PhD in theoretical physics at University of Chicago, my father was asked to be part of some odd study that he never explained to me. He said he refused to do it because the two men who spoke to him about it insisted he could never tell his family if he did it. Sometimes I wonder if he was part of that study anyway, without his knowledge. That's because his friend at Argonne National Lab took special interest in human mood and cognition, and also took special interest in my father.

My dad's friend, as far as I could tell, was a good person. But he had collaborated in the 1970s with Carl Pfeiffer, one of the researchers involved with MK-ULTRA. In that program, and in the follow-up MK-SEARCH program, Pfeiffer had tested drugs (including LSD) on prison inmates. However, my dad's friend's collaboration could have been entirely innocent, since it occurred after the end of MK-SEARCH. This post-MK-SEARCH collaboration apparently focused on healing mental illness among violent criminals, using vitamins and minerals.

My dad's friend volunteered to help inmates at the local prison — he started a chess club for prisoners, with the help of many volunteers from Argonne. I still wonder if he was involved in some kind of extension of MK-SEARCH that was kept secret, perhaps even from him. I also wonder if Argonne physicists and other personnel were part of these tests as well.

In any case, it might be interesting to do a FOIA request on any Dept. of Energy research on inmates at the Stateville Penitentiary in Illinois, during the years 1960-1980. I don't know if anything would turn up, but it's something to consider pursuing for those who would like to understand more about MK-ULTRA and

MK-SEARCH. At this point, I don't consider myself one of those people.

In Fall of 2023 a few weeks after I'd moved from Northern California to the DC area, I had a dream that got me thinking about one potential explanation for all of this. I've had many dreams that seem to be precognitive or telepathic, but this is the only dream that ever gave me a working phone number. I've copied the dream description directly from my dream journal — and printed it out for those who can't read my early-morning handwriting.

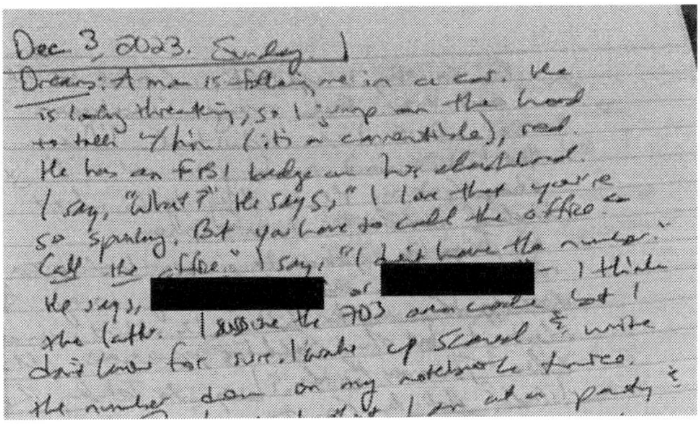
Dec 3 2023. Sunday
Dream: A man is following me in a car. He
is looking threatening, so I jump on the hood
to talk w/ him (it's a convertible), red.
He has an FBI badge on his dashboard.
I say, "What?" He says, "I love that you're
so spunky. But you have to call the office.
Call the office." I say, "I don't have the number."
He says, [redacted] or [redacted] — I think
the latter. I suppose the 703 area code but I
don't know for sure. I wake up scared & write
the number down on my notebook twice.

Dream journal entry.

"Dec 3, 2023. Sunday.

Dream: A man is following me in a car. He is looking threatening, so I jump on the hood to talk with him (it's a convertible), red. He has an FBI badge on his dashboard. I say, 'What?' He says, 'I love that you're so spunky. But you have to call the office. CALL THE OFFICE.' I say, 'I don't have the number.' He says, 'XXX-XXXX' or 'XYX-XXXX' — I think the latter [the second number was one digit different from the first]. I assume the 703 area code, but I don't know for sure. I wake up scared and write the number down on my notebook twice."

⊗ ⊙ ⊡

Of course after I had breakfast, I searched for the second number. Nothing in the 703 area code, so I tried Washington DC and Bethesda, two nearby area codes with plenty of government offices (202 and 301). They both led to the same organization — the 202 area code plus the XYX-XXXX number was for the organization's headquarters in the 80s, and perhaps the 90s. The 301 area code plus the same XYX-XXXX number was for that organization's headquarters in the 2000s and 2010s, which had apparently moved to Bethesda. The organization is responsible for assessing and protecting against damage from radiation. I am not giving its name here because I suspect the dream was just leading me to a conclusion by using a related phone number, and I don't think it's kind to the current employees to put the phone number here.

The first website that showed up when I searched for the number was actually a document: https://tinyurl.com/RadiationDoc. It's June 1978 memo from Hartman B. Mowery of the Defense Nuclear Agency, with the subject line "Ionizing Radiation Studies." It's about developing a uniform approach to managing the effects of exposure to low-level ionizing radiation. The memo stresses this is a "people problem" and they are not concerned with animal or environmental effects (implying that human participants would be the only appropriate participants).

In case the link I included above ever fails to work, I've included the first page of that document so you get the basic point.

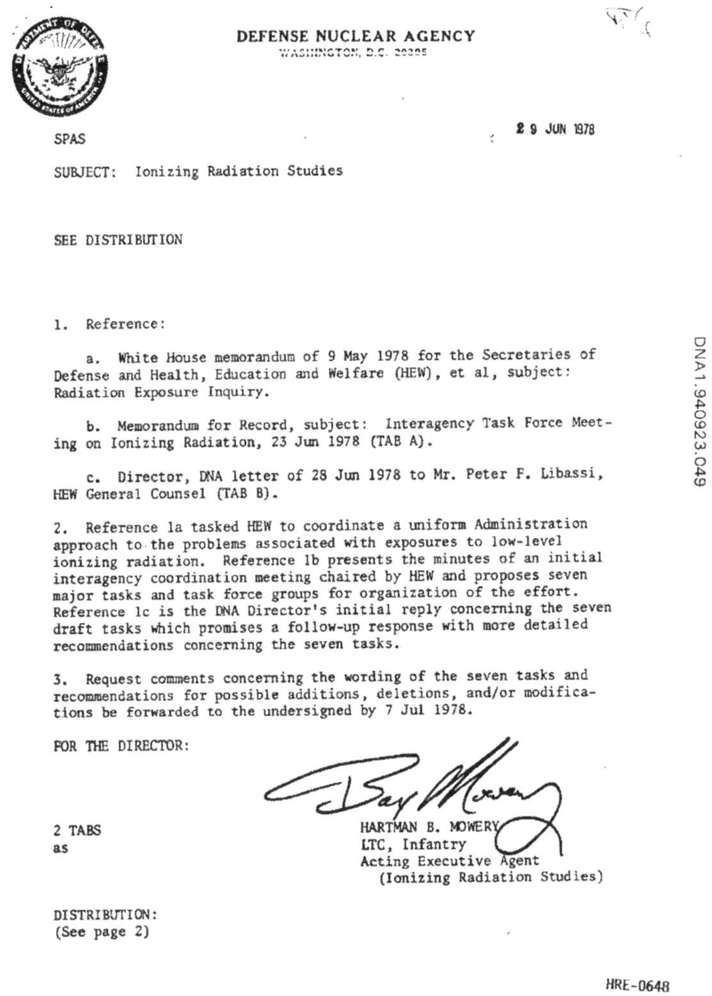

DEFENSE NUCLEAR AGENCY
WASHINGTON, D.C. 20305

SPAS

29 JUN 1978

SUBJECT: Ionizing Radiation Studies

SEE DISTRIBUTION

1. Reference:

a. White House memorandum of 9 May 1978 for the Secretaries of Defense and Health, Education and Welfare (HEW), et al, subject: Radiation Exposure Inquiry.

b. Memorandum for Record, subject: Interagency Task Force Meeting on Ionizing Radiation, 23 Jun 1978 (TAB A).

c. Director, DNA letter of 28 Jun 1978 to Mr. Peter F. Libassi, HEW General Counsel (TAB B).

2. Reference 1a tasked HEW to coordinate a uniform Administration approach to the problems associated with exposures to low-level ionizing radiation. Reference 1b presents the minutes of an initial interagency coordination meeting chaired by HEW and proposes seven major tasks and task force groups for organization of the effort. Reference 1c is the DNA Director's initial reply concerning the seven draft tasks which promises a follow-up response with more detailed recommendations concerning the seven tasks.

3. Request comments concerning the wording of the seven tasks and recommendations for possible additions, deletions, and/or modifications be forwarded to the undersigned by 7 Jul 1978.

FOR THE DIRECTOR:

2 TABS
as

HARTMAN B. MOWERY
LTC, Infantry
Acting Executive Agent
(Ionizing Radiation Studies)

DISTRIBUTION:
(See page 2)

HRE-0648

DNA1.940923.049

First page of document at the location of the first search result for the dream phone number.

Super intriguing, but (again) it could all be coincidence. Certainly my dreaming about a phone number does not prove anything.

However, I have replicable scientific evidence of my own ability to retrieve information from some kind of informational field (from experiments not performed by myself). So I read the document with interest, since I figured the dream sent me there.

I didn't have the courage to call the dream phone number until I was writing this part of the book, just a few days ago (around Sept 1, 2025). I asked the administrator who answered the phone if they might want a copy of my new book — she obligingly asked me what it was about.

I told her it was essentially about finding your own truth, respecting that, acknowledging that you matter, and moving into unconditional love for yourself and others. She and I giggled a little as I described the book, because we both knew this agency was probably not the right audience. She said something like, "Well, you can send it, but I can't promise anyone will read it." I suspect she's right.

Over the almost two years between that dream and now, I started to think about the radiation exposure connections. Again, it could all be a coincidence — but there were a few more of which I have only recently become aware. Some of the architects of SOAR seem to have had a clear relationship with radiation exposure. And perhaps other SOAR students (aside from me) did too.

My SOAR teacher, with whom my mom had worked as a learning disabilities instructor at Copeland School in Libertyville, had a husband who was a drill sergeant near Aiken, SC, which is near the Savannah River National Lab, which processed radioactive materials. The creator of SOAR in Aiken, SC (Joyce Ross) had a husband who was a chemical engineer for that same lab. And the original poster of that Reddit thread talked about being a SOAR student in Nevada, famously a nuclear test area.

I was in college and grad school in the 1980s and 90s, but if I had been a psychologist or neuroscientist working in a federal lab back then, I think I would have felt compelled to understand the creative, psychic, and intellectual abilities/disabilities of US persons exposed to radiation. I would like to think I wouldn't have performed research without consent — but I would still wonder if there could be intergenerational effects of radiation exposure.

I might think we could ethically study kids in the 80s whose parents had high exposure in the 40s, and study kids in the 90s and later whose parents had high exposure in the 50s (to ionizing radiation) or the 60s-80s (to electromagnetic radiation).

Maybe I'd see if I could help restore some of these kids to better functioning, trying to balance out the harm that was done by nuclear bomb testing and other sources of radiation. A chemical could be mixed with some kind of pink fruit syrup and be given to kids who were thought to be at risk of radiation poisoning, and we could track perceptual and cognitive performance over time. Certainly I could foresee legal and reputational concern about this research becoming public, so I imagine that the research would be classified. But still I would like to think that basic consent for the studies would be asked of parents, even if the results and analytical details were not revealed.

Where we go from here.

I want to be clear (again) that all these speculations are based on my own experiences, public records, and my expertise. I do not have the answers. The motivations for these programs I believe to be a complex mix of well-intended and thoughtless ones. Regardless, we know one clear truth:

It is always unethical to experiment on anyone, especially children and protected populations like people with disabilities or prisoners, without informed parent/guardian consent and meaningful participant assent.

If someone ever tells you experimenting on protected populations is necessary for national security, let them know that there must be another way to do whatever it is that needs to be done other than putting already fragile people at risk without their informed consent and assent.

I want to speak to others out there who may be in various stages of the recovery process from this kind of (or other) unethical actions. You may have had, like me, childhood abuse and other traumas that you were and are also dealing with — and it might feel like this kind of "re-litigating the past" is the least of your worries. You may feel shame or fear about your secrets, and you are right to wait until you feel safe before you disclose anything publicly.

Once you feel safe enough to speak your truth, please go slowly and lovingly towards speaking out within the context of your commitments, clearances (if any), and compassion. The resistance and fear in the rest of the world is huge, and you can go at your own speed and proceed in your own way. You can write a song, do a piece of art, post a post on social media.

Get a care team you trust — and make sure it includes you. Consult your team.

People outside your care team might be critical. They may say things like, "You have an outsized sense of your own importance. You don't matter that much for your experience to make a difference." But that's a great lie — in fact, it's *the* great lie.

I believe you matter so much that G-d gave you a body, a mind, and a soul. That's three complex things that have to work together across dimensions throughout your lifetime. That's a lot of effort for Love to invest into someone who doesn't matter.

So we are lying to ourselves and each other when we say that we don't matter. That lie allows all sorts of secrets to stay closed off inside us. And it allows shame to fester. Releasing shame is one of the most powerful acts we can do to support our wellbeing.

My husband does a variety show for adults called "As You Are." One of the things he does in the show is to encourage people to hum their secrets. People love it. One by one they'll come up and think of a secret, then they'll hum it into a megaphone. Afterwards, the look on their faces is the best.

Where I'm at now is that I don't need to know some kind of externally declared truth about my past to move forward in health and love. I believe several earnest and talented teachers worked hard to support my thinking and creativity. Maybe they were wives of military servicemen who knew that some damage might have been done in the past, and they wanted to support children in becoming the best they could be. Maybe they were simply good educators who had nothing to do with any government program. Maybe they were tracking a hybrid human-alien program, or they were intelligence officers working on a super-soldier program. Here's another idea — our future selves have time-jumped (or UFO-jumped) back to the past to let us know we needed to fix our radiation-damaged genetics as a species. Who knows? I don't.

My best bet is these educators and study designers were well-meaning people. Probably some of them were keeping secrets they felt were necessary to keep. Everyone was doing the best they could, as usual. As always.

Everyone involved deserves forgiveness and love, even if they had bad intent at the time. It took me some time to get here, but this is where I am.

What can I do now? As I move forward and do work in the world, I know that I can find a way to tell the truths that matter. I can work hard to access Universal Love to support our world in letting go of unnecessary secrets. As a scientist who works with both children and adults to understand psychic capacities and extraordinary human functioning, I can do ethical experiments with informed consent and meaningful assent. And I can join the rest of those lucky enough to have survived a world of too many secrets by reaffirming that your body, your soul, and your mind matter.

We are all doing the best we can.

MY UNIVERSAL TRUTHS

Disclosure implies something that was hidden is now revealed. But there are some universal truths that are more powerful than any real or imaginary international cartel or deep state. These truths transcend time, authority, and physical bodies. They are the opposite of hidden; they are more like bedrock.

I think of these universal truths as the source of what I call the "Inspiracy," and I think you've already heard them in some form in this book and hopefully many other places.

1. Universal Love is a central truth that can never steer your soul wrong. Accessing Universal Love takes courage, but it grows your soul in ways that also support other people in growing theirs. You get to keep the growth in your soul beyond the death of your physical body.

2. It is easy to get caught up in the desire for material power and influence. Trying to get access to material power and influence feels like a smart way to behave, because they can actually make your life feel better. But these pursuits don't help the evolution of your soul — except to teach us that they are not the point, which is a helpful thing to learn.

3. Secrets are an attempt to keep ourselves and those we love

safe. The motivation for secrets arises from the fear that if a person or group of people knew the secret, bad things would happen. This is a normal fear, because we all have a drive to help our physical bodies survive and thrive. But there is a mistaken belief that supports secret-having and secret-keeping. It's the belief that there is not an informational substrate or a wise universe of information that allows us to access all the secrets as well as the perceived reasons for keeping them. This mistaken belief creates the misunderstanding that a secret can actually be kept successfully over space and time.

4. Many of us have a very reasonable fear of losing material power/influence/health/life. To feel better temporarily, we can get in the habit of using conditional love to shape others' behavior — "You're in this group/relationship if you obey these rules and have these beliefs, you're out if you don't." This conditional love is often confused with Universal Love, but Universal Love has no constraints around who gets to feel loved. This is why unconditional love feels like a threat when we forget that our souls cannot be nourished by material power and influence. When we forget this, we can forgive ourselves and each other by reminding ourselves of the transcendence available in Universal Love.

5. Four soul-food groups nourish the soul. Feeding yourself from these four groups every day will make your soul healthy and strong, regardless of your material health/influence/power. They are:

    - connecting with Universal Love/G-d,
    - hearing/speaking the truth,

- learning goodness,
- feeling grateful.

6. When souls all over the world and everywhere in the universe connect with love to each other, when we try our best to support and nourish each and every soul with the four soul-food groups, when we access universal love most of the time — that will be really great fun!

Interpretation Ideas

This is not a scholarly book. If you find problems with the facts, consider these problems a gift to support your wise discernment. Also, there are many ideas in this book smashed in together. I just write like that.

I hope you can use your wise discernment to grasp the meaning that works within you over time.

None of the ideas here are mine — none belong to me. Or anyone else. All have been thought of before, just like each word in this book has been used many times. But I do like the way these particular words go together, and I was the human-communication tool system sub-part that put them together this way, so that's why I'm calling it my book.

If you figure out the symbolic language, let me know. I drew these symbols with a silver-colored marker on my brown-paper-bag dust jacket for a middle-school English textbook in my SOAR class back in 1982. I don't really know what they mean. I know they are used to express only ideas of the soul. They can't be used to refer to material things at all.

I have some other glimmerings about them:

1. Each symbol is an idea or concept, not a letter or a word.

2. They always occur in groups of three.

3. The symbols describe present moment only, and they read from right to left.

4. The first symbol (on the right) refers to what is happening with G-d/Universal Love.

5. The second symbol (in the middle) refers to what is happening simultaneously with the human soul.

6. The final symbol (on the left) refers to what is happening simultaneously with the relationship between Universal Love and the human soul.

7. The symbols can contain each other to communicate more complex happenings in each of the three positions.

8. Every symbol has a circle or square on the outside, and there is often a dot or line(s) on the inside.

9. There can be one or two lines on the inside, but if there are two lines, they are always perpendicular.

10. Lines can be diagonal, horizontal, or vertical, and they always connect with the outline of the circle or square that holds them.

11. The symbols can be thought of as 2-D representations of 3-D objects.

12. You can increase interpretational possibilities by mentally building a "donut" of 3-D versions of the three symbols in each grouping, each one morphing gently or suddenly into the next symbol.

I hope together we'll understand what the groupings revealed in this book mean. More importantly, I hope we let them do their important work in us, even if we don't understand them.

Thank you for your patience, openness and love. It's somewhat of a pain to do the work to read this book all the way through, and I very much appreciate you doing that work.

If you are interested in learning more about the topics in this book and my work, please visit TheInspiracy.love.

Gratitude

I feel deep gratitude for all my teachers, students, collaborators, and co-authors, all my mentors and mentees, and every friend I've ever had or will have in the future. There are so many times I've looked at the acknowledgements of a book and realized it becomes a long list of names, with little meaning conveyed. I started making a long list of names, and then I wondered if some folks maybe wouldn't want to be thanked in such a strange book. Or maybe I should only use their first names? Then I gave up.

What if I forgot someone? Or included one person with no last name ("Thanks Dan!") and one of my friends thought it was another Dan, when it was actually All The Dans. So I want to be clear — I am grateful for everyone I've ever liked, loved, or disliked and struggled to love, including my family (the greatest teachers of all).

You are magnificent, you are well loved, and I am grateful for you. And yes, I did include you. Did you notice?

Also, because I know I can name these two names, my husband Brooks Palmer and my son Joseph Mossbridge are my two most precious spacetime anchors. I love and appreciate them with all my heart, in all available dimensions.

And G-d, of course, encompasses and lives in all of us. There is no greater gift than to realize that.

Made in the USA
Las Vegas, NV
23 May 2026

4ee5b0ce-5bdb-466b-915a-b84d77c31ca3R01